Patterns in Music

by Gloria Chen

A pattern repeats over and over again. You hear patterns in music.

clap! clap! stomp!

You can make patterns, too. Try this pattern. Clap, clap, stomp.

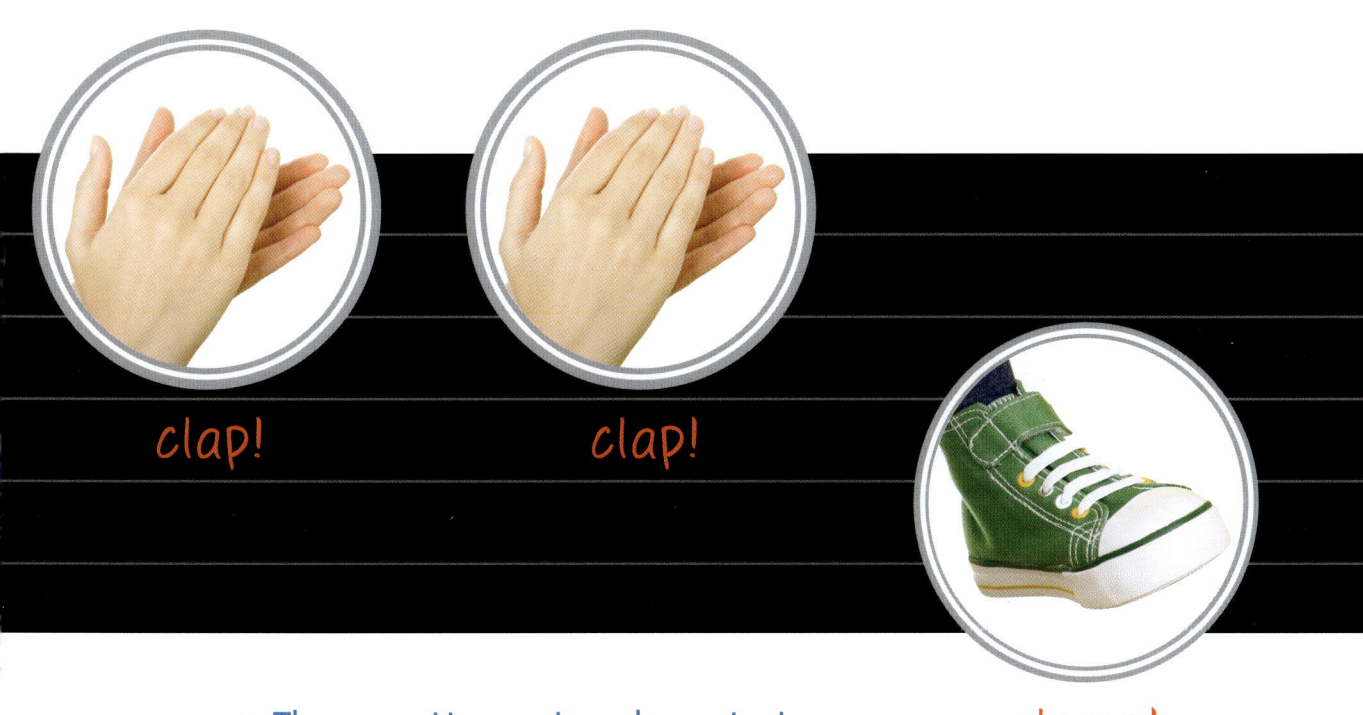

clap!   clap!

▲ The pattern is clap twice, then stomp.

stomp!

You can use instruments to help you make patterns.

ting!

stomp!   stomp!   stomp!

Try this pattern. Stomp, stomp, stomp, then hit the triangle.

ting!

stomp!   stomp!   stomp!

▲ These pictures show the pattern.

5

A drum is another instrument that you can use to make a pattern.

tap

tap

clap!

First tap the drum. Next tap the drum again. Then clap.

tap   tap   clap!

▲ You can repeat the pattern by tapping the drum, tapping the drum, and clapping.

You can make a pattern
with this instrument, too.

shake

shake-shake   shake-shake

The pattern is shake one time,
next shake two times, then shake
two more times.

shake

shake-shake   shake-shake

▲ The maraca is a fun instrument to play.

Some instruments can make high sounds. The violin can make a high sound.

The tuba makes a low sound.
You can make a pattern of high
and low sounds.

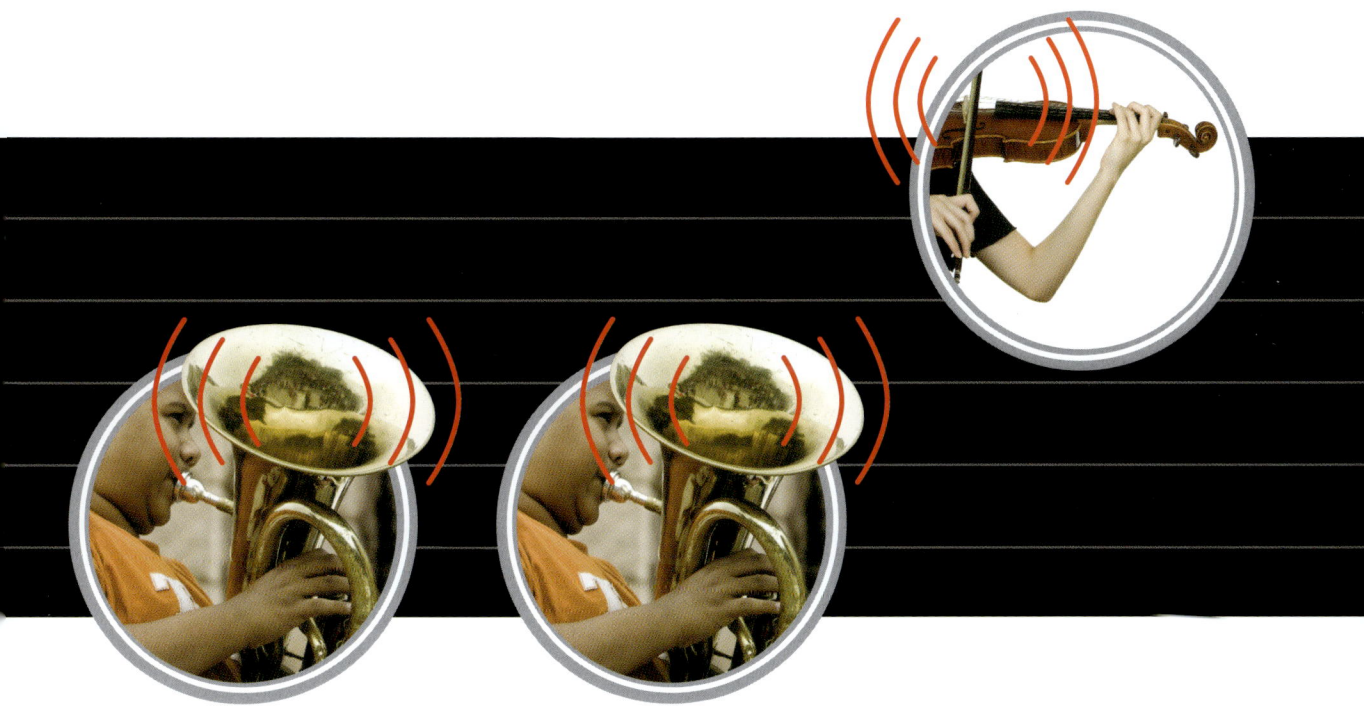

▲ The pattern is low, low, high.

If you blow into a tuba one long time, you make a long, low sound.

If you blow into a tuba one short time, you make a short, low sound.

▲ You can make different sounds with the tuba.

You can use a tuba to make
a pattern of long and short sounds.

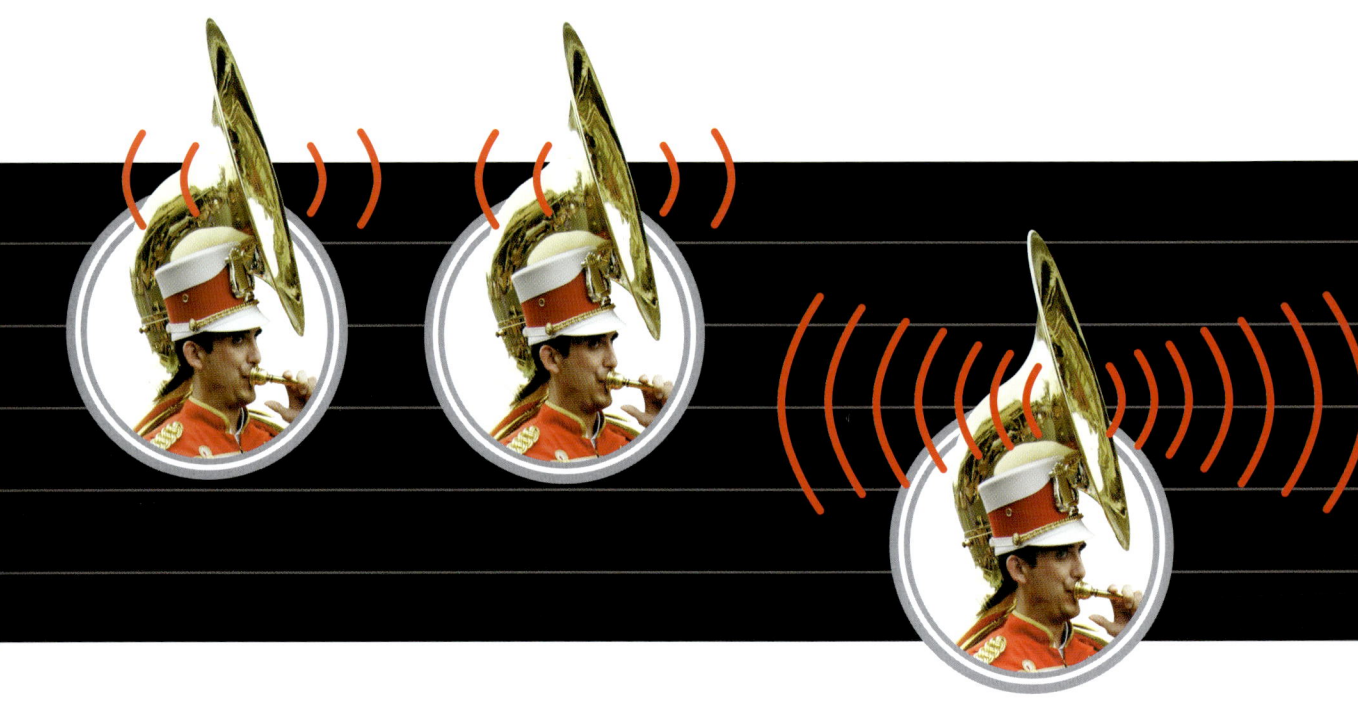

The pattern is short, short, long.
Then repeat short, short, long again.

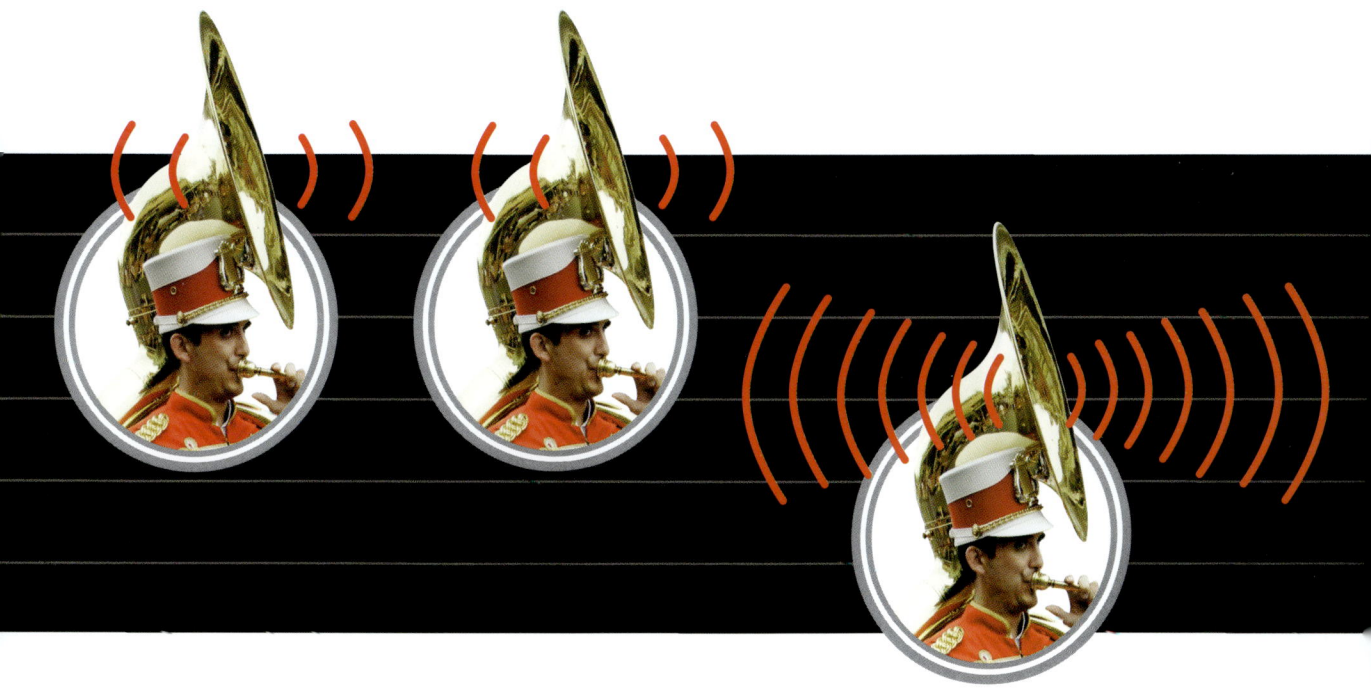

▲ This pattern repeats itself over and over again.

You can find patterns in music.
You can create patterns, too.